Acknowledgements

KV-575-708

CIRIA and the project participants wish to acknowledge the contributions to the project and this report of Professor David Gann and Dr Ammon Salter of the Science Policy Research Unit at the University of Sussex, especially for the material summarised in Section 1.

CIRIA is grateful to the participants in this project for their support, good advice, hard work and funding.

Participants

Mr Ron Williams (chairman)	Mott MacDonald
Dr Tim Broyd	WS Atkins
Mr David Brocklehurst	HJT Ltd now WSP Group plc
Mr Peter Deason	Kvaerner Technology
Mr Ken Farrer	Montgomery Watson
Mr Steven Groák/Mr Barry Austin	Ove Arup & Partners
Mr Roger Jones	Hyder
Mr Peter McGee	Amec
Mr Andrew McNaughton	Balfour Beatty
Mr Ed Plewa	Maunsell Ltd
Dr Michael Starr	Halcrow
Mr Richard Thomas	Posford Duvivier

Research team

Dr Ghazwa M Alwani-Starr	CIRIA
Dr Peter Bransby	CIRIA
Prof David Gann	SPRU at University of Sussex
Mr Ammon Salter	SPRU at University of Sussex

DEDICATION

This report is dedicated to the late Steven Groák of Ove Arup & Partners.

CIRIA C531

London, 2000

The management of technical excellence in design organisations

Ghazwa Alwani-Starr
BSc PhD CEng MIStructE

ing best practice

n SW1P 3AU
20 7222 1708

Summary

This report is aimed at those managing design teams in construction companies of all types and who are interested in promoting the concept of technical excellence within their organisations.

It describes CIRIA's investigations of the ways in which some of the UK's largest construction companies manage technical excellence. While concentrating on the conclusions drawn and recommendations for good practice, this publication also provides background information on practice in other industries and will aid the reader in putting current UK construction practice in context.

Twelve of the UK's leading design consultants and contractors took part in the study, which was carried out between January and September 1998. The aims were to establish how the participating organisations managed technical excellence and to investigate the environment within which technical excellence flourishes in design organisations.

A confidential (and necessarily lengthy) report containing all the raw data collected during the project was initially prepared for participants. This summary report is for wider distribution and application, and has been produced with permission from the participating organisations. It aims to introduce both the concept of technical excellence and some practices that construction companies could adopt to promote technical excellence within their organisations.

Management of technical excellence within design organisations

Alwani-Starr, G

Construction Industry Research and Information Association

Publication C531   © CIRIA 2000   ISBN 0 86017 531 6

Keywords		
Technical excellence, benchmarking, business improvement, design organisations		

Reader interest	Classification	
Design company managers, construction managers, project managers, government advisers and academics	Availability	Unrestricted
	Content	Survey results and guidance
	Status	Committee-guided
	User	Construction sector managers, academics and government advisers

P
693
ALW

Contents

1 Introduction

One area in which design service providers compete is their specialist technical knowledge and professional competence. The management of this 'technical excellence' in design organisations is therefore an important short-term competitive issue, as well as being vital for the long-term survival and growth of the firm in an increasingly competitive worldwide environment.

The need for work in this area was highlighted by one of CIRIA's largest Core member companies in 1997. Discussions with other member companies revealed that consultants and contractors have developed a variety of practices, both informal and explicit, to manage the function of technical excellence. There also appeared to be a wide variation between (and perhaps within) companies. The benchmarking of practice across major consultants and contractors was therefore likely to bring benefit to all, as well as directing management attention to the important topic of developing and monitoring technical excellence.

This report is in four sections. Section 1 offers an introduction to the study and background information on innovation management theory and practice in other industries. Section 2 describes the study in some detail, while Section 3 presents the results. The final section of the report offers a commentary on the results and suggests areas of improvement that may be relevant to the wider construction industry community.

1.1 BACKGROUND

Management of technical excellence in design organisations depends on creating an environment that allows good design to flourish. Attempts to define 'design' often fail because it involves a variety of conceptual and physical attributes that do not fit inside narrow boundaries. The management of design is complicated both by this variety and by the many interactions between a range of professional and engineering disciplines.

As Steven Groák suggested, little is known about ways of managing design but a great deal is known about creating an environment which allows good design to emerge. He suggested that design flourishes in environments which motivate individuals to be creative, to generate new ideas, and to take chances (Groák, 1992). Such environments allow for an element of risk taking, such as that associated with innovation, and for failure. Environments promoting design excellence enable individuals to draw on external ideas, to communicate their ideas and to share ideas with others. Design organisations need to find the right mix of organisational routines to sustain these environments and motivate their employees.

The management of technical excellence for design organisations involves providing organisations and individuals with the opportunity to develop to their full potential. At times this might involve routine support and information services, and at other times it may involve ensuring creativity. In this respect, management needs vary from the routine to the exceptional. Routine support may involve the development of information sources, such as on-line CD-ROMs or the maintenance of physical libraries. Exceptional support may involve bringing in external speakers to spark discussions among key staff in the organisation.

1.1.1 Innovation

The management of technical excellence is shaped by an organisation's ability to use the resources of both the organisation itself and outside organisations for the stimulation and production of new ideas, technologies and practices. Technical excellence is dynamic, evolving over time and it arises through an organisation's ability to sustain innovation.

Innovation refers to new technical designs, manufacturing techniques, management and commercial activities that are successfully commercialised. The Department of Trade and Industry defines innovation as the "successful exploitation of new ideas". It includes large-scale changes in the technological state-of-the-art and small-scale changes in technical know-how (Tidd *et al*, 1997; Freeman and Soete, 1997).

Innovation can be embodied in new products and processes. Product innovation involves the development of new products for commercialisation, whereas process innovation refers to the development of new techniques for production. Product innovation is the commonly recognised form, but it is by no means the most important. Both forms of innovation play a role in stimulating growth for firms. In fact, there is often a tight connection between the development of new products and new processes, as new products give rise to new processes or vice versa.

Patterns of innovation differ across industries, firms and sectors. Different sectors have specific characteristics and activities, for example the automobile sector is dependent on scale economies. These differences lead to many types of innovation process and heterogeneous outcomes. Results of innovation processes are often unknown until other future events have occurred. In areas where there is limited knowledge about current practices, it is hard to see the benefits (or costs) of organisational innovations or new technologies. Innovation therefore occurs in an uncertain environment.

Although innovation processes are uncertain and heterogeneous in outcome (see Section 2.1.3), there are important ways to manage them. Organisations develop routines for managing change which become embedded into their structure, yet some routines are better than others. Although it is possible to learn from other organisations, it is difficult to copy their routines as they are built over time. No simple model of best practice exists for innovation and organisations thus have to find their own ways of doing things.

1.1.2 Innovation in design organisations

Design organisations are involved in the development, production and use of complex products and systems. These are high-cost, engineering-intensive products and systems which embody large numbers of customised, interactive sub-assemblies and components, including embedded software (see Miller *et al*, 1995; Hobday, 1998).

They include:

- many products in the built environment such as 'intelligent buildings' – hospitals, R&D laboratories and education establishments, control stations, office, retail and distribution complexes, factories and military establishments (Gann, 1992)
- utilities and communications infrastructure
- power, oil and gas, offshore and process plants
- transport nodes and infrastructure.

Complex products and systems are usually produced through one-off or very small batch processes, often involving temporary coalitions of firms. In some respects, these facilities are never quite completed because of their long life, with often unpredictable changes in patterns of use. Indeed some are in almost continual cycles of change. This creates a dynamic environment for design in which the uses and costs of construction need to be balanced against future needs, cost in use and lifecycle costs.

Management of design is complicated by the discontinuous nature of project-based production processes in which there are often broken learning and feedback loops. The choice of technology and management of innovation takes place under conditions where it is not usually feasible to test full-scale prototypes. Simulation and modelling is therefore of great importance in front-end decision making, planning and execution. Product definition, development, simulation, testing and production usually entail the transfer of knowledge within complex networks of suppliers, involving a large number of interactions between many different specialists. This includes the need to deal with technical decisions in which the interdependency between components and subsystems creates the need for an exchange of technical know-how across a range of professional and engineering disciplines.

2 The study

The desktop study and literature review that was carried out gave an insight into the theories behind innovation, learning and the management of technical excellence. They enabled the shifting of participants focus from day-to-day project and customer service delivery to the wider context of technical excellence delivery, encompassing all elements of an organisation.

2.1 IDEA DEVELOPMENT

In addition to the desktop study and literature review, pilot interviews were held with two members of the participating group and an interview was conducted in Japan. The information gathered highlighted three main areas for discussion and the need for clarification to be addressed by the study.

2.1.1 Factors affecting the development of a definition

For the participating group, definitions of what constitutes technical excellence depended upon who is using the term and why they are using it. The definitions of "technical excellence" include:

- a formal, written mission statement or marketing tool developed and owned by the group, division or operating unit – trading units managed autonomously, sometimes defining their own terms of reference for technical excellence

- a formal, written part of company procedure, eg QA

- an informal personal belief of practice or aspiration, based on core competencies held by individual practitioners. This definition varies between practitioners, is unwritten and relates closely to tacit knowledge learnt through custom and practice on the job.

Perceptions of technical excellence traditionally relate to academic notions of finding the most elegant and theoretically perfect engineering or design solution. This led to a definition in which engineering prowess was given primacy over other considerations such as aesthetics, buildability and fitness for purpose. Engineers practising in this mode have sometimes assumed the role of supreme expert and gained a reputation for arrogance, based on their technically sublime but sometimes unnecessarily sophisticated solutions.

Two ways of defining technical excellence emerged from the pilot interviews.

1. Definitions based on internal performance criteria and individual motivations:

   → technical excellence is "the ability to do the work you tell others you can do, and do it effectively and efficiently"

   → technical excellence is "based on personal interests developed by individuals".

2. Definitions based on an outward-looking, market-facing approach relating to the company's reputation:

   → "it is important to think of the technical requirements of the customer and not whizz-kiddery"

   → "customers want smart results, but technical excellence is represented in the capability to break down complex issues into simple items"

   → "customers want value engineering and innovation skills and they view delivery mechanisms as part of a total package which involves excellence in practice. Customers may well perceive delivery mechanisms to be as important as the product being delivered. Technical excellence therefore embodies organisational and management capabilities"

   → "commercial success and client satisfaction demonstrate the ability to perform with technical excellence".

A view from Japan indicates that leading Japanese engineering design and construction companies take a bottom-up approach to defining technical excellence. This is embedded within their performance improvement and quality system known as QSCD, standing for Quality, Safety, Cost, Delivery (Environment and Moral have also being added by some companies).

The onus for performance improvement is with the individual engineers and project managers in each firm. All staff in the larger firms know the principles of QSCD, but it is up to individuals to define the terms in their own way. For this reason, it is not possible to easily measure performance improvements or benchmark practices because ways of working change from one engineer or project manager to the next.

Technical excellence is therefore an instinctive and implicit part of Japanese engineering and construction culture. This does not mean that technical excellence becomes subsumed beneath other more explicit activities and targets. Japanese engineers and project managers compete proudly within their companies to demonstrate their capabilities in a spirit of continuous improvement.

2.1.2 Factors affecting the management of technical excellence

The desk study, pilot interviews and debate by the research team led to the factors affecting the management of technical excellence being classified under eight headings.

1. The size, ownership and financial robustness of the company

The ownership, control and financial robustness of a company has a profound affect on the attitude and confidence of managers, and hence on their approach to managing technical excellence. Willingness, or ability, to commit to the investment that will lead to a virtuous circle of enhanced technical excellence and an advantageous market position will depend on the financial robustness of the company. It is hard for those struggling to survive to break out from a spiral of decline.

The size of a company, the number of locations it occupies and whether or not they are widely dispersed are also relevant to the management of technical excellence. This aspect has particular impact on the communication systems within an organisation. The nature in which developments and innovations, whether project-based, firm-based or external, are communicated to staff has a major impact on how they are implemented, evaluated and improved for the benefit of the firm.

2. **Culture**

A company's recognition of the need for innovation and continuous improvement influences the actions it takes to facilitate the management of technical excellence. Is there an expectation of individuals and teams deliberately seeking to learn from their own or others' experience? A touchstone could be the openness of the company to the sharing of its own knowledge with others, including competitors, as a basis for a more rapid learning and development of competence.

3. **Structure**

The necessity of separating larger companies into a number of operational units has the potential of creating significant organisational barriers, which will inhibit the development of an integrated set of competencies. Perhaps one of the most powerful tools in influencing behaviour will be the use of appraisal criteria for the performance of individuals responsible for the different operational units. If there are obvious incentives in working collaboratively with other units and explicit encouragement to develop technical excellence within the unit, these are likely to be powerful influences. But the question remains as to how performance on each of these two criteria could be judged.

4. **Company strategy**

Whether or not a firm reinforces its commitment to technical excellence by having a clearly articulated company strategy, including an intention to invest in this area, has a major impact on a firm's management of technical excellence. Successful strategies include a definition of responsibilities for technical excellence and clear allocation of these responsibilities, perhaps with a nominated director taking the lead. Strategies also include communication and delegation plans covering all levels of the organisation.

The same considerations apply for the funding of the development of technical excellence and whether budgets are a corporate or local responsibility. The advantage of giving the manager of an operational unit the freedom to commit the resources they believe to be appropriate to the technical excellence function is that any expenditure will be directed to their market needs. On the other hand, they may well choose to invest less (or more) than is judged appropriate from the corporate perspective.

5. **Market and work characteristics**

From an internal perspective, the scope and diversity of the areas of work covered by a firm will affect how easy it is to develop an outstanding competence: large "one-stop shop" organisations have the more challenging task than small specialists. The company's organisational structure, whether by region, customer, technical discipline or otherwise, will be relevant.

This is heavily influenced by the external environment, which is changing the characteristics of work packages that firms undertake. The average size of job that a firm undertakes is becoming smaller and, coupled with this, control of the quality of the overall product is not falling within its boundaries (see Section 4.1.3). Hence it is becoming less clear whether a firm has produced a technically excellent solution to a problem.

6. **Knowledge sharing and networking**

The issue of how knowledge is shared across the organisation is obviously central to the development of a corporate competence. Networking, both within the company and externally, as a source of intelligence and information is likely to be a significant component of the management of technical excellence. How this is best done is unclear, especially now that electronic networking is becoming universal. Nonetheless, given the information overload that most individuals experience, it is likely that personal face-to-face networking will retain its value, especially in terms of testing and assessing the value of information received. The increasing importance attached to the subject of knowledge management is relevant here. Given the nature of construction, it is likely that companies will wish to develop their knowledge base in partnership with others in the supply chain and even with clients. The current emphasis on PFI (Public Finance Initiative) and partnering both point in that direction.

7. **Marketing**

Part of the marketing of individual companies will be the promotion of their own technical excellence and the management of their reputation. The techniques used to convey technical excellence by different companies are likely to vary but will include published papers, contributions to conferences, newsletters, awards and other promotional activities. It would perhaps be helpful if there were clearly understood and objective measures of technical excellence as that would allow companies to focus their energies and be more convincing to clients when competing with organisations which are less conscious of technical excellence issues.

8. **Customers and clients**

The customers served by a company will clearly influence the way firms do business and, for example, the ways in which design solutions are produced. The crucial link is with the customers' perceptions of value and the ways in which technical excellence can contribute to this.

The question then is how technical excellence can be managed in such a way as to maximise the perceived value by the customer. This may require communicating the process of delivering the product; the sophisticated customer or client is likely to want demonstrated value engineering and innovation skills to be sure that they are receiving the best service possible. The growing diversity of markets and the new drivers, such as PFI, mean that companies have to develop their technical excellence faster than in the past. The balance between flexibility to acquire new skills and strength in traditional disciplines is therefore likely to be an issue.

2.1.3 Impact on the research

The participating group thought that, in one sense, technical excellence can only be judged by the quality of the product. But such a view is unhelpful in examining the ways in which performance may be improved. The group then agreed that delivered technical excellence results from a combination of actions and inputs – a process. A successful process can be defined as one that delivers appropriate levels of technical excellence (for example, performance, safety, quality of the finished product) within time and resource limitations. Better management of the process therefore has the potential to deliver improvements in time, cost or delivered quality. The focus needs to be on the scope for improvement of the process of delivery of, rather than on the delivered, technical excellence.

The definition adopted was thus as follows: the management of technical excellence is the management of a set of strategic and operational processes that together lead to the delivery of appropriate levels of performance and quality within time and resource limitations.

It is clear from the definition and the issues highlighted above that managing and improving technical excellence is very complex. However, given the time and financial limits imposed in the project, it was important to select issues for further examination where benchmarking could lead to real improvements. Suggestions of such issues were:

- attitudes and approaches to technical excellence – company policy, strategy statements, responsibilities for technical excellence and staff expectations
- the management of separate operating units
- communications systems for the capturing of project knowledge and innovations – individual versus company knowledge
- the production of tangible outputs, eg published reports, papers, contributions to seminars and conferences
- training and recruitment policy
- use of IT
- marketing strategy
- information sources accessed by the firm and external links
- staff
- approach to and investment in R&D.

These issues were discussed by the participants during their first meeting and this debate led to the formulation of a framework for the project that served until its end. The participants agreed that the constituent elements of the main processes that impacted on the management of technical excellence were as listed below, grouped together under generic headings. The groupings were made on the understanding that some of the factors may be placed under more than one heading, depending on the particular aspect being considered. It must also be noted that all factors relate to and impact upon each other.

The development of shared values, strategy and policy which promote technical excellence.

An examination of the following areas of company practices was conducted:

- company mission statement
- business strategy and line responsibility
- company strategy for continuous improvement
- strategy for "getting it right first time"
- strategy for continuous learning and knowledge sharing
- resource sharing policies.

The effective selection and development of staff.

Here, the key elements of staff selection, development and training were examined:

- staff qualifications and skills at recruitment
- mapping required core competencies – recruitment and projects
- training and development policy
- staff retention
- staff appraisals
- awareness of staff resource and mobility policy
- secondment
- motivation
- team building.

The applications of those procedures or frameworks which allow staff to perform at their best.

The key procedures that were thought to influence staff performance were examined:

- post-project evaluation
- review procedures and capturing experience
- utilisation of QA procedures
- staff incentive schemes
- procedures to encourage continuous improvement.

The use of knowledge systems which allow the sharing and build-up of knowledge.
Here, the key elements of knowledge systems that, when well developed and managed, allow the sharing and build-up of knowledge were examined:

- library
- intranet
- access to public databases
- in-house project database
- peer review
- technical support systems – both internal and external
- learning networks.

Once this framework was established, data on practice within participating organisations was collected through a number of iterations. Four rounds of data collection were undertaken, and each round of data collection deepened the thinking and knowledge of the issues surrounding the management of technical excellence.

2.2 THE RESEARCH PROCESS

The research was carried out in the following stages:

- pilot interviews and desk study to explore the general issues
- preparation of a brief paper on issues arising
- meeting of the participating group to discuss issues and decide on areas in which comparative data would be desirable
- collection of data on company practices
- two-day workshop involving the participating group and the research team to debate current company practices
- further data collection
- one-day workshop to debate findings and highlight areas where further data is required
- further data collection
- collation of data and preparation of final project output
- agreement of project output and further work.

This process was developed in recognition of the value of the data that exists within the participating organisations. As can be seen, the data was collected in three stages, each stage deepening the survey and enriching the knowledge. After each stage, participants were given the opportunity to meet and debate the results in an open workshop format where a deeper understanding of the issues was facilitated by the research team.

2.3 GROUP DEVELOPMENT

It is important in the context of the study to make a brief comment on the development of the participating group. The group consisted of individuals who held positions of responsibility within their organisations. All were directors and senior managers and most had extensive experience of managing design. The first meeting established the ground rules for work. The group was keen to exchange knowledge and experience and agreed that information on actual current practice would be exchanged so as to be of maximum benefit to all concerned.

After initial hesitation regarding the confidentiality of some information, the group developed a relationship that enabled frank and open discussions which were found to be extremely useful by participants. This developed because:

- the group found that no single organisation had solved all the issues relating to the management of technical excellence. All organisations were facing the same challenges and issues and they were all at different stages of finding solutions

- there was a great deal to be gained from frank and open exchange of information and very little to be gained from attending meetings where few real or valuable contributions were made

- the group was protected at every stage of data collection. CIRIA presented all information in a non-attributable manner until all members of the group agreed to share openly. This provided a safety net at every stage which enabled participants to keep their contributions confidential if they wished to.

Towards the end of the project, the group had developed a relationship which they valued. The group agreed that they would like to continue to work together to address other areas of their business where open debate would be useful.

Due to the lack of hard numerical data that could be collected in the context of the project, and because no single participating company had an edge over others in all aspects, the study moved away from traditional benchmarking into open comparison and debate. Adopting this approach enabled members of the group to gain a much deeper, and hence much more useful, insight into others' practices as the information gathered was not limited by the narrow interpretation of numbers alone.

3   Company practices

3.1   SHARED VALUES, STRATEGY AND POLICY

An examination of the strategies adopted by the participating organisations for developing and communicating company-wide values and policies revealed a variety of approaches. These ranged from having a detailed mission statement, including targets on issues such as health and safety, cost and time, to those which had a general and more inspirational approach to the mission, for example as articulated in the key speech delivered by Ove Arup in 1970 which is seen to be still relevant today. Only half of the companies taking part in the study had a mission statement that mentioned technical excellence. However eight out of twelve focused on the delivery of excellent customer service. It was also interesting to note that while some organisations had long-established and understood mission statements, others had developed their mission statement very recently. On both sides, however, there were doubts on how well the mission statement was communicated and understood throughout the organisations.

Responsibility for the implementation of the company mission was invariably trusted to division or group directors. Only one company systematically reviewed its business strategy in light of its mission on an annual basis. The review was communicated to staff, who had the opportunity to discuss it and give feedback.

Data was collected on how the companies monitored their performance with regard to moving towards achieving their shared values and mission. Although all respondents felt that they were indeed progressing, few had in place explicit measurement or mission review procedures. Those who did have such procedures collected client feedback and reviewed profit or growth targets in an attempt to gain an insight into their performance. No organisation collected information on the cost of implementing its mission statement. The indicators used here ranged from spend on Investors In People initiatives, R&D and training to spend on marketing or business development.

In terms of having a strategy for continuous improvement, eight out of the twelve participating organisations gave information. Only one organisation had an explicit commitment to continuous improvement which was supported at board level. Other companies' approaches were through supporting a number of internal initiatives which aimed at improving particular aspects of the business. In addition a number of firms viewed their quality assurance system as the main route to causing practice improvements. Five out of nine firms felt that being involved in formal R&D contributed towards continuous improvement and seven out of nine had established innovation or similar newsletters in an attempt to institute a culture of continuous improvement.

Further information was gathered on R&D (see Section 2.2). Firms taking part in the study carry out R&D as part of projects and spend is consequently booked to projects. This is a vital observation about the way that construction companies operate. As spend in time and money on R&D is generally booked to projects, it is often lost as a resource in that figures do not appear separately. Hence questions such as 'what did we learn?' and 'how can we make sure we tell everyone else, so that they don't spend money learning the same lessons?' do not often get asked. Not surprisingly, therefore, the results showed that some organisations spent ten times more than others as a percentage of their turnover on R&D.

The company that had an explicit policy for continuous improvement supported at board level also had the philosophy of 'getting it right first time' at the heart of its quality system. None of the other firms had an explicit commitment; they also found it difficult to single out a particular strategy element that aimed at getting it right first time. Most firms relied on project reviews through quality, health and safety and other management systems to improve the company's ability to achieving 'getting it right first time'.

The participating companies' approaches to learning and knowledge sharing ranged from seeking to achieve Investors in People standards to staff encouragement in obtaining professional and other qualifications. Involvement with the work of research and academic organisations, together with involvement in networks such as CIRIA's Construction Productivity Network which facilitates the sharing and increasing of knowledge, was also recognised by most participants as valuable input towards continuous company improvement.

On the question of resource sharing, most organisations viewed their resources as being mobile and accessible to all parts of the organisation. Problems were highlighted, however, in developing systems whereby information on staff skills and company competencies may be accessed quickly and easily. Staff mobility was also affected by protective attitudes of some department or group leaders, who were reluctant to let good people be transferred to other divisions or work on other projects.

It is worth mentioning here that in some organisations (not any of those involved in the study) an attempt has been made to resolve the issue of reluctance to share knowledge and/or staff. This has been achieved by developing a system whereby knowledge and staff may be traded. In other words, an employee has to contribute to the development of the company's knowledge base before being able to gain access into it. Staff members may be exchanged on the basis of the skills they possess. For example, a member of staff knowledgeable in IT project management tools may be exchanged for one knowledgeable in health and safety management tools.

An examination was undertaken of the current and future importance of all these issues to the firms taking part in the study. It was interesting to note that all firms saw the various elements of company mission and policy as important, but predicted that they will become much more important in the future.

3.2 STAFF SELECTION AND DEVELOPMENT

In general, participants in the study found it much easier to provide information on their practices in this area than in the first area of study. Most of the organisations taking part had very detailed and specific procedures for staff selection and development. Training programmes were in place, with some intensive induction courses implemented for a number of years. Staff turnover was in the region of six to seven per cent, which was viewed as being acceptable.

Although assessment of staff qualifications and skills at recruitment was extensive, it was often based on a detailed job description rather than individual competencies in light of required company competencies. In other words, rarely were company competencies mapped and recruitment then based on obvious gaps or desired requirements. However, eight out of the twelve participating organisations had a skills database. Most organisations reported very limited use of this facility due to a lack of reliability of the data and overall maintenance of the database.

A variety of recruitment methods were used, ranging from traditional advertising and interview by technical staff to the employment of headhunters for more senior positions. The use of psychometric testing to assess the personality of potential employees was becoming more widespread, together with the use of specialist recruitment companies and specialists in interviewing techniques and facilitation.

The use of facilitators was more widespread where team-building exercises were used (eight out of twelve). Although some organisations rarely used formal team-building techniques (except where specified by a client), some had established exercises and varied informal social activities as key elements in company team building.

The performance of teams was assessed either through financial and programme-linked achievement of targets or through subjective and informal assessment. Some comparisons of team performances were undertaken internally but, where these happened, financial performance was at the heart of the assessment.

All the organisations carried out staff appraisals. The elements of the appraisal ranged from assessment of technical performance, training and development in the last twelve months to assessment of relationships with colleagues and career aspirations. The relationship between appraisals and planned career progression was less clear. Some appraisals fed directly into a development plan; others had no formal relationship with career planning at all. This prompted the research team to investigate the reasons why participating organisations carried out appraisals. A range of responses was gathered. Some identified the main reason was for the development of a training plan, while one firm used appraisals to assess pay and bonus awards.

Most organisations aimed to appraise all their staff annually, and most achieved 70–100 per cent. Problems due to shortness of available time were encountered with appraising all staff annually. One firm had found a solution to this by developing an appraisal system that is relative to the seniority and level of experience of the appraisee. Under that system, senior staff were appraised at least once every three years while more junior staff were appraised at least annually.

All participating firms had excellent intentions as far as training was concerned. Few achieved their target spend in terms of money or time. Average total time spend on training was in the region of 1.6 per cent of total man-hours, when most of the participating organisations had a target of nearer four per cent which they had not been able to achieve.

Issues relating to general staff matters were judged by participants to become increasingly important to their business in the future. Less emphasis was being placed on technical development.

3.3 PROCEDURES

The participating group considered post- project evaluations and review procedures to be two of the most important factors in the management of technical excellence. It was therefore interesting to note that only six of the twelve participating organisations carried out post-project evaluations on a regular basis. Where they were carried out they were directly linked to the QA system in operation. Three of the six recognised the importance of feedback and had procedures in place to ensure that lessons are communicated internally.

Eight out of the twelve participants had some form of review or procedure in place for capturing experience, including reviews of the financial performance of projects. Five out of twelve had a form of communication of the findings of the review.

It is interesting to note here that almost 100 per cent of offices of all participating organisations were certified to ISO 9001.

An examination of design checks and design reviews was also carried out. The study showed that although 100 per cent of projects underwent a detailed design check, only about thirty per cent underwent a design review. This suggests that there is potential for significant savings if the industry was to review design decisions more frequently. During the study, the participating group did not make explicit the connection between value management and value engineering to reviews. It is thought that if use of these concepts had been surveyed, an even smaller percentage of projects undergoing design reviews would have been found.

Eight of the participating group of companies described their project QA procedures. These procedures were being applied to all projects, major and minor, and were rigorously audited. One company was undertaking a major review of its procedures brought about by changes in its practices and the way it did its business. For the eight organisations, QA seemed to provide the basis on which most project level procedures were developed.

In relation to the perceived increasing value of staff retention and development, five out of twelve participants had some form of incentive scheme. Three out of the five were related to pay in the form of a profit-sharing or bonus scheme. One had an innovation award, another a reward for commercial performance.

To cover the procedural side of the general issue of continuous improvement, the group agreed to consider how client feedback and staff feedback are collected and acted upon. Six out of eleven organisations had a proactive procedure for collecting client feedback. Some of these were in general terms, on average every eighteen months, while others were in relation to specific projects and were either continuous throughout the life of a project or only on completion. On average, between five and ten per cent of clients were asked for feedback and the results were generally communicated to senior managers so they could address the issues which arose.

Half of the firms taking part in the study actively collected staff feedback through a variety of routes. Departmental meetings, staff councils and suggestions schemes were in operation. One firm had an incentive scheme (financial award) in operation, associated with their suggestion scheme.

Mechanisms for capturing, sharing and using lessons from experience were seen as the most important aspect for improving the future performance of firms. This is discussed further below.

3.4 KNOWLEDGE SYSTEMS

All the participating firms had extensive traditional libraries which were valued as an important company resource.

Most companies had developed in-house databases. These were mainly to record project details and performance, particularly of a financial nature, and were commonly used for marketing and sales information. Often, they were not systematically maintained.

Other sources of information for internal technical support and exchange of knowledge included internal/regional technical seminars, annual technical meetings, feedback databases (embryonic), internal lectures and seminars by internal and external experts, and speakers and activity in the research community. Where undertaken, peer review to promote the sharing of knowledge was conducted as part of QA systems.

Technical support was therefore very prominent in many of the organisations, with a strong technical development programme in operation in most. Learning networks, as a method for sharing and development of knowledge, were in various forms of operation in most organisations. Nine of the participating firms encouraged the development of internal learning networks and one company reported a variety of innovative approaches of sharing knowledge such as the development of a 'dating agency' and 'expert switchboard' to provide internal technical support effectively (see Section 4.1.3).

Nearly half of the firms had recently established intranets for the communication and sharing of information. Some debate was taking place within most organisations on the type of information to post on their intranets. Some contained purely marketing and corporate information, while others contained detailed technical information and data. The use of the intranets was not being monitored regularly at the time of the survey.

Respondents felt that the use of the Internet and intranets will increase substantially in the future and will have a larger impact on company profitability.

Spend on IT was a difficult figure to quote by participating firms as some had a central IT budget while others had local budgets. The difference in spend was also a reflection of the different positions in the replacement/upgrading cycle which the organisations taking part in this study occupied. Thus in cases where intranets or new PCs were being installed, a higher spend was obviously observed. The range of spend on IT was between a half and seven per cent of turnover.

The area of capturing, storage, maintenance and access to information was seen by the participants to be a key factor in influencing productivity and competitiveness in the future.

4 Discussion and conclusions

4.1 IMPACT OF THE STUDY ON THE MANAGEMENT OF TECHNICAL EXCELLENCE IN DESIGN ORGANISATIONS

4.1.1 Corporate issues

Perhaps the most important observation that can be made regarding the management of technical excellence is the extent to which it is governed by the management of the whole range of processes and procedures that together make an organisation function. There is no one answer that applies to all organisations and different areas of the business will have a different impact on technical excellence in different organisations.

The most important action that firms need to implement at a corporate level to encourage technical excellence is to demonstrate a commitment to excellence. Managers need to examine their organisations in detail, analyse the skills of their key staff, and consider the procedures they have put in place and the support systems they have available. This analysis should examine how the company's mission and values are transferred throughout the organisation – staff, procedures and systems – to ensure the delivery of the expected levels of performance.

Whatever a company does to demonstrate commitment requires investment. This investment could be in various forms:

- in time, eg for directors to address staff regularly

- in money, eg in developing effective communication systems, or

- in people, eg making sure that staff with the correct skills are recruited, trained, motivated and retained.

The firms participating in the study are organised in a number of ways. They varied from small close-knit practices to large multi-disciplinary, multinational corporations. The way in which they were organised had little impact on their activities as far as the management of technical excellence was concerned. The impact that the size and organisation of a firm had was a function of how effective the communication mechanisms were in these firms and the performance criteria used.

It is not a simple task to take a snapshot of an organisation and examine its staff, processes and procedures. These have grown out of years of changing management styles, technologies and people. The changes are, of course, continuous as companies adapt to a constantly changing trading environment. It is therefore not surprising that the management of technical excellence at a corporate level is not dependent on particular organisational structures or management systems, but rather on paying adequate attention to the key issues highlighted above: commitment, investment, communication and measurement.

But what can companies do tomorrow to improve their practices?

4.1.2 Staff issues

Project participants felt that one of the most influential factors on the delivery of technical excellence is staff selection at all levels: company, division, section and project team. Effective mechanisms that ensure the right calibre of staff, with the required skills and motivation, are placed at every level of the organisation were seen as essential.

Again, this is not a simple task. It was recognised by the participants that difficulties in recruitment are widespread. Two reasons for these difficulties are highlighted.

1. There is a shortage of candidates of the required calibre applying for jobs with construction companies. This is a major issue for the construction industry as a whole and it demands a shift in perception of the value of an engineer. Companies, however, can attract better candidates than competitors by developing attractive staff development and reward packages.

2. Construction companies' assessment criteria for applicants are becoming less technical and emphasis is shifting towards personal and 'soft' skills. Historically, the recruitment processes in operation in most firms are based on the assessment of technical knowledge. It is now recognised that technical managers responsible for recruitment require training in new interview and candidate assessment techniques.

Companies that want staff with particular skills are having to invest heavily in development and training.

There is also the issue of identifying key staff within an organisation. The study showed that, in any organisation, two levels of staff have most influence on company and project performance: manager and project team leader. The manager was critical because he or she set the performance measures and hence is the driving force behind the outcome. The team or project leader was seen as critical because he or she is in charge of delivery and hence in charge of ensuring the right people, plans, programmes and other delivery mechanisms are in place to deliver the required outcome. It is essential for organisations to assess who in their organisations has the capacity to influence and, therefore, how to allocate the responsibility for ensuring that overall corporate targets are met.

Finally, construction is not an individual activity. Working in a team is vital internally and is being applied more extensively, with the integration of design and construction, externally. Working in teams and being able to communicate with all levels of the supply chain will be an everyday activity for most construction professionals. Firms wanting to give their businesses a competitive advantage ought to consider methods for the development of teamwork and communication within their organisations.

4.1.3 Technical issues

The real understanding of client requirements and the effective translation of these requirements into a design brief have long been major challenges for designers. In the context of management of technical excellence, however, the effective preparation of a client's brief was seen by the participants as being vital for ensuring the delivery of technical excellence.

The process does not stop there though. The management of design so that solutions are achieved that fully meet the client requirements sounds straightforward, but it is not easy to achieve in practice. With the move towards the integration of design and construction, the start and end of any design activity has become blurred. With this, responsibility for design has become blurred and now, more than ever, communication throughout the supply chain has become vital. Interfaces between professions and professionals need to be communicated and agreed at the earliest stages of a scheme. (Refer to CIRIA's FR/CP/78, *Managing design in civil engineering design and build*, for further discussion of this issue.)

In technical areas such as construction, learning from mistakes through making use of feedback and reviews is an essential element in the delivery of technical excellence. It allows the build-up of knowledge and the development of companies' know-how. Whether through project reviews, staff appraisals or client surveys, the effective communication and use of feedback was a major issue for participating firms.

It was not a question of little or no feedback, but in fact the opposite: firms appeared to spend considerable resources on the collection of feedback. However, important questions were often not asked, such as:

- why are we collecting this information, in this format, from these people?
- what are we going to use this information for?
- who are we going to communicate it to?
- what are they going to do with it?

Most participating firms identified the effective use of feedback as an essential, organisation-wide issue that needed to be addressed.

4.1.4 Communication issues

Communication is the word most commonly highlighted in this section. Effective communication within firms was seen as the major driver for the delivery of technical excellence. Communication of corporate thinking and strategy, communication of learning and new knowledge, communication of feedback and lessons learnt are all essential elements in the effective conduct of business. In design organisations this is vital. Vast amounts of information and knowledge need to be managed and communicated, sometimes internationally, very quickly.

The combination of effective management and communication of knowledge and information, together with the effective harnessing of available technology, is the future key to the delivery of excellence at all levels.

4.2 SCOPE FOR IMPROVEMENT – ISSUES FOR LEADING DESIGN ORGANISATIONS

In addition to reflecting on the overall delivery of technical excellence in design, individual participants were asked to indicate, privately, what they thought were the areas which their own organisations needed to develop in order to deliver technical excellence.

The following areas of improvement were highlighted:

- understanding the concept of technical excellence and developing a culture of excellence
- developing and implementing effective feedback and knowledge transfer systems
- developing effective staff selection procedures, and mapping, maintaining and developing staff and their competencies as a resource
- understanding customer needs, developing effective briefing procedures and providing adequate solutions

- managing design more effectively
- developing effective project review procedures
- developing strategies that promote team working
- improving communication up, down and across the firm, particularly on motivational/soft issues
- developing effective, meaningful and useful performance measures.

Implicit in some of the above points is the management of design activities and the performance measurement of design. The latter was the subject selected by participants for continuing to work together. Work in this area became more important to the business of the participating organisations as government and other client bodies published new procurement guidelines, which stress the importance of integrated design and construction and the need to demonstrate value for money.

Performance measurement of design activities is the subject currently under study in the second stage of this research project. It is due for completion by the summer of 2000 and a summary report, similar to this, will be published by the end of 2000.

4.3 AND FINALLY...

Although participants were a little cautious at the beginning of this project, at its conclusion they had no doubt as to the value of taking part.

The study highlighted the value both of learning how to improve specific company practices but also the value, in general, of sharing information between companies, debating practices and networking.

Perceptions of what should be treated as a confidential company policy, system or procedure were changed and the importance of learning from each other, communicating openly and harnessing all sources of available knowledge were those that emerged as more important.

Collaboration, partnerships and the effective harnessing of all available knowledge by the right people are likely to be at the heart of future competitiveness.

References and further reading

Bergen, S A (1990) *R&D management: managing projects and new products*, Basil Blackwell, Oxford

CIRIA (1998) *Standardisation and pre-assembly – adding value to construction projects*, Report 176, CIRIA, London

Construction Task Force (1998) *Rethinking construction – report of the construction task force*, Department of the Environment, Transport and the Regions, London

Davies, A C (1997) "Lifecycle of a complex product system", *International Journal of Innovation Management*, Vol 1, No 3, pp 229–256

Freeman, C and Soete, L (1997) *The economics of industrial innovation*, 3rd edition, Pinter Press, London

Gann, D M (1992) *Intelligent buildings: producers and users*, SPRU, University of Sussex, and Electrical Contractors' Association, London.

Gann, D M, Hansen, K, Bloomfield, D, Blundell, D, Crotty, R, Groák, S and Jarrett, N (1996) *Information technology decision support in the construction industry: current developments and use in the United States*, Department of Trade and Industry overseas science and technology expert mission visit report, SPRU, Brighton

Gann, D M and Salter, A (1998) "Learning and innovation management in project-based, service-enhanced firms" submitted to *International Journal of Innovation Management*

Groák, S (1992) *The idea of building*, E&FN Spon, London

Hobday, M (1998) "Product complexity, innovation and industrial organisation", *Research Policy*, Vol 26, pp 689–710

Leonard-Barton, D (1995) *The wellsprings of knowledge*, Harvard Business School Press, Boston, Mass.

Miller, R, Hobday, M, Leroux-Demers, T and Olleros, X (1995) "Innovation in complex systems industries: the case of flight simulation", *Industrial and Corporate Change*, Vol 4, No 2, pp 363–400

Patel, P and Pavitt, K (1997) "The technological competencies of the world's largest firms: complex and path dependent, but not much variety", *Research Policy*, Vol 26, No 2, pp 141–156

Prencipe, A (1997) "Technological competencies and products' evolutionary dynamics: a case study from the aero-engine industry", *Research Policy*, Vol 25, No 8, pp 1261–1276

Rosenberg, N (1994) *Exploring the black box. Technology economics and history*, Cambridge University Press

Tidd, J, Bessant, J and Pavitt, K (1997) *Managing innovation: integrating technological, market and organisational change*, John Wiley & Sons, Chichester

Womack, J P, Jones, D T and Roos, D (1990) *The machine that changed the world*, Maxwell Macmillan International, New York